Discovering

Life on a Coral Reef

by Melvin and Gilda Berger

SCHOLASTIC INC.

ISBN 978-0-545-35173-7

12 11 10 9 8 7 6 5 4 3   13 14 15 16 17 18/0

Printed in the U.S.A.   40
First printing, May 2013

Photo Credits: Photo Research: Alan Gottlieb

Cover: © age fotostock/SuperStock; Back cover: © Mark Conlin/Alamy; Title page: © Georgette Douwma/
Nature Picture Library; page 3: © Carlos Villoch-MagicSea.com/Alamy; page 4: © WaterFrame/Alamy;
page 5: © Carlos Villoch-MagicSea.com/Alamy; page 6: © Darryl Leniuk/Getty Images (RF); page 7: ©
Cavan Images/Getty Images (RF); page 8: © Jeff Hunter/Getty Images; page 9: © Franco Banfi/Getty
Images; page 10: © Mark Smith/Science Source/Photo Researchers; page 11: © Krzysztof Odziomek/
Shutterstock; page 12: © Chris Newbert/Minden Pictures; page 13: © Michael Pitts/Nature Picture
Library; page 14: © Ian Scott/Shutterstock; page 15: © Tim Laman/Getty Images; page 16: © Jeff Hunter
/Getty Images

Coral reefs form in warm seas.

Reefs are made of shells from tiny animals.

Is this reef near the shore?

Reefs grow as the shells pile up.

Corals are different colors.

Corals are also different shapes.

Many kinds of animals live around a reef.

Plants live there, too.

Is the clown fish orange and white?

Clown fish swim near coral reefs.

Lionfish live there, too.

Sponges sway in the water.

Do the eels blend in with the coral?

Moray eels hide among the coral.

Where is the shark's mouth?

Reef sharks look for fish to eat.

Parrot fish eat the coral.

sk Yourself

1. Where do coral reefs form?
2. What are reefs made of?
3. Do reefs have animals and plants living around them?
4. Can you name one kind of reef fish?
5. What do reef sharks eat?

You can find the answers in this book.